Preface

Welcome to insects. Insects are found all over the world. We live in nature. Nature contains many insects. Insects are very beautiful. Many flowers attract insects. Insects are fly away from one flower to another flower. They help to pollination. Insect live in nest. The scientific study of insects are called entomology.

This book contains colourfull pictures of varities of insects. The pictures are taken from pixabay. All pictures credit goes to pixabay.

Dedication

To My Parents

Ant

Aphids

Antlions

Bees

Beetles

Butterflies

Bugs

Cockroaches

Damselfly

Dragonfly

Earwig

Grasshoppers

Lacewings

Lice

Mantis

Mayfly

Moth

Scorpion

Stick Insect

Termites

True bugs

Wasp

References

1. "A To Z of Insects." *A To Z of Insects - Amateur Entomologists' Society (AES)*, https://www.amentsoc.org/insects/fact-files/a-to-z-of-insects.html.

2. "A To Z of Insects List - General Knowledge for Kids: Mocomi." *Mocomi Kids*, 2019, https://mocomi.com/a-to-z-of-insects/.

About the Author

Anupam Rajak received his B.Sc in Botany from the Raghunathpur College, Sidho-Kanho-Birsha University. He has published several articles in international journal. His email address is anupamrajak1234@gmail.com

www.ingramcontent.com/pod-product-compliance
Lightning Source LLC
Chambersburg PA
CBHW051837210526
45473CB00005B/1918